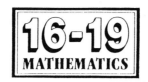

Differential equations

Unit guide

The School Mathematics Project

CAMBRIDGE
UNIVERSITY PRESS

Main authors	Stan Dolan
	Mike Leach
	Tim Lewis
	Richard Peacock
	Jeff Searle
	Phil Wood
Team leader	Jeff Searle
Project director	Stan Dolan

The authors would like to give special thanks to Ann White for her help in preparing this book for publication.

Published by the Press Syndicate of the University of Cambridge
The Pitt Building, Trumpington Street, Cambridge CB2 1RP
40 West 20th Street, New York, NY 10011-4211, USA
10 Stamford Road, Oakleigh, Victoria 3166, Australia

First published 1992
Reprinted 1996

Produced by 16-19 Mathematics, Southampton

Printed in Great Britain by Watkiss Studios Ltd., Biggleswade, Beds.

ISBN 0 521 42660 X

Contents

Introduction to the unit
(for the teacher)

This unit should be attempted towards the end of the 16-19 Mathematics course, after a substantial amount of calculus has been covered. In particular, the differential equations chapter of *Mathematical methods* is an essential prerequisite.

It is recognised that many students working through this unit may be doing so without the benefit of substantial contact time with a teacher. The unit has therefore been written to facilitate 'supported self-study'. It is assumed that even a minimal allocation of teacher time will allow contact at the start and end of each chapter and so

- solutions to all thinking points and exercises are in the students' texts;

- a substantial discussion point in one of the opening sections enables the teacher to introduce each chapter;

- a special tutorial sheet can be used to focus discussion at a final tutorial on the work of the chapter.

After studying this unit, students should appreciate that a range of practical and theoretical problems can be investigated by both numerical and symbolic techniques. The use of computers or graphic calculators is essential to a successful development of the material and their use should be encouraged throughout.

Chapter 1

Students are reminded of their earlier experiences of differential equations. In particular, the technique of finding an approximate solution by the numerical step-by-step method is revised.

Chapter 2

This chapter develops the use of the step-by-step method for simultaneous first order equations. In tackling the discussion point, students should be encouraged to acquire and develop programming skills, so that these are readily available for the rest of the chapter. Second order linear differential equations are solved by reduction to two simultaneous first order equations and higher order equations are tackled in a similar way.

The *Solution sketcher* program, on the 16-19 Mathematics disc *Real functions and graphs*, can be used for differential equations of the form $\frac{dy}{dx} = f(x, y)$. Solution curves for other differential equations can be sketched by plotting the numerical results by hand. Alternatively, software such as David Tall's *Graphical calculus III* (Glentop) can be used.

Chapter 3

At the start of this chapter, the student should be in a position to solve any linear differential equation by a numerical technique, but should also be aware that there can be limitations in the accuracy and validity of such solutions.

The discussion point in this chapter should emphasise the potential power of a symbolic solution as well as justifying the technique of separating the variables.

Section 3.2 introduces the important ideas of complementary functions and particular integrals for linear differential equations. Students should aim for a thorough understanding prior to the further development of this method in Chapter 4. The solution of first order linear equations using integrating factors is covered on an extension tasksheet.

Chapter 4

The solution to second order linear equations with constant coefficients is developed using complementary functions and particular integrals. The need for two boundary conditions in order to obtain a particular solution should be stressed at the discussion point. The auxiliary equation approach to obtaining a complementary function is developed, and, once students have become familiar with the techniques, the method is justified. The cases of repeated roots and imaginary roots are dealt with and students should be reminded of the appropriate section of *Complex numbers* if necessary.

The unit ends with an extended exercise, covering a range of real problems. This should emphasise the modelling potential of differential equations which was discussed in Chapter 1.

Tasksheets

1 Review

1.1 The order of a differential equation

(a) How do differential equations arise?

(b) Why is the solution of differential equations important?

(c) What methods of solution are available?

(a) Differential equations involve rates of change and are found in many fields of study, ranging from science and engineering to business and commerce and from pure mathematics to health and medicine.

Whenever quantities are changing, the mathematical equations which model these quantities are likely to involve rates of change. Examples of some standard models from a range of applications are as follows:

$\dfrac{dN}{dt} = -kN$	Radioactive decay
$\dfrac{d^2\theta}{dt^2} = -\dfrac{g}{l}\sin\theta$	The motion of a simple pendulum
$\dfrac{dP}{dt} = rP(1-kP)$	Population growth
$-\dfrac{p}{n}\dfrac{dn}{dp} = k$	Elasticity of demand
$\dfrac{dv}{dt} = g(1-kv^2)$	A body falling in a resistive medium
$\dfrac{dT}{dV} + \lambda T = 0$	The adiabatic gas law
$\dfrac{dx}{dt} = k(a-x)$	Wilhelm's law for chemical reactions
$\dfrac{d^4y}{dx^4} = k$	Vertical deflection of a horizontal beam

You may be able to think of examples from other subject areas.

(b) For a differential equation which models some particular situation, the solution is used to attempt to predict the value of one of the variables when the other(s) are altered. For example, the solution might show how the size of a population varies with time or might show the safe loads for a structure such as a bridge.

(c) The techniques available for the solution of differential equations are essentially of two types, numerical and symbolic. Both types of method are developed further in this unit, which concentrates on the solution of differential equations rather than on their formulation.

Numerical solutions

REVIEW

COMMENTARY

TASKSHEET 1

1.　(a)

x	0	0.2	0.4	0.6	0.8	1.0
Step 0.2　y	0	0.20	0.40	0.61	0.86	1.16
Step 0.1　y	0	0.20	0.40	0.62	0.88	1.20
Step 0.05　y	0	0.20	0.40	0.63	0.89	1.23

(b)　The exact solution is $y = \frac{1}{4}x^4 + x$ and so $y = 1.25$ when $x = 1$. At $x = 1$, the errors are:

Step	Error (to 2 d.p.)
0.2	0.09
0.1	0.05
0.05	0.02

In this example, the error appears to be approximately halved as the step size is halved.

(c)

x	1.0	1.2	1.4	1.6	1.8	2.0
Step 0.2　y	1.25	1.65	2.20	2.94	3.96	5.33
Step 0.1　y	1.25	1.68	2.28	3.09	4.19	5.66
Step 0.05　y	1.25	1.70	2.32	3.16	4.30	5.83

At $x = 2$, the exact solution is $y = 6$. Again, the error appears to be approximately halved as the step size is halved.

The errors are much larger on $1 \leq x \leq 2$ than on $0 \leq x \leq 1$. The sketch of $y = \frac{1}{4}x^4 + x$ shows the graph to be more curved on $1 \leq x \leq 2$.

2.　(a)

t	$\frac{\pi}{12}$	$\frac{\pi}{8}$	$\frac{\pi}{6}$	$\frac{5\pi}{24}$	$\frac{\pi}{4}$
Step $\frac{\pi}{24}$　h	−0.47	−0.29	−0.04	0.22	0.46
Step $\frac{\pi}{48}$　h	−0.47	−0.27	−0.02	0.24	0.47

(continued)

10

(b)   The exact solution is $h = -\frac{2}{3} \cos 3\,t$.

t	$\frac{\pi}{12}$	$\frac{\pi}{8}$	$\frac{\pi}{6}$	$\frac{5\pi}{24}$	$\frac{\pi}{4}$
h	-0.47	-0.26	-0.00	0.26	0.47

Both numerical solutions are very good approximations to the exact solution.

(c)

	t	$-\frac{\pi}{12}$	$-\frac{\pi}{24}$	0	$\frac{\pi}{24}$	$\frac{\pi}{12}$
Step $\frac{\pi}{24}$	h	-0.47	-0.66	-0.76	-0.76	-0.66
Step $\frac{\pi}{48}$	h	-0.47	-0.64	-0.71	-0.69	-0.56
exact	h	-0.47	-0.62	-0.67	-0.62	-0.47

The errors using the smaller step length are about half those obtained when using the larger one.

The errors in this domain are significantly larger than for $\frac{\pi}{12} \le t \le \frac{\pi}{4}$. This is because the gradient of the exact solution, $h = -\frac{2}{3} \cos 3t$, changes rapidly in $-\frac{\pi}{12} \le t \le \frac{\pi}{12}$.

Tutorial sheet

1.   (a)   (i) $120x$ (ii) $\sin x$

       (b)   (i)   All derivatives higher than the fifth are zero.

            (ii)   The fourth derivative of $\sin x$ is $\sin x$ and so the derivatives simply repeat the sequence $\sin x, \cos x, -\sin x, -\cos x$.

2.

$$y = \sin wx$$

$$\Rightarrow \frac{dy}{dx} = w \cos wx$$

$$\Rightarrow \frac{d^2y}{dx^2} = -w^2 \sin wx$$

$$= -w^2 y$$

3.   (a) $f(x) = x^2 - 1$

$$\Rightarrow f'(x) = 2x$$
$$\Rightarrow f'(x) - 3x\ f(x) = 2x - 3x\ (x^2 - 1)$$
$$= 5x - 3x^3$$

       (b) $g(t) = e^{-2t} \Rightarrow g'(t) = -2e^{-2t} \Rightarrow g''(t) = 4e^{-2t}$

           $g''(t) + 3g'(t) + 2g(t) = 4e^{-2t} + 3(-2e^{-2t}) + 2e^{-2t} = 0$

4.   (a) $\frac{dh}{dt} = 8 - 10t$ gives the velocity of the ball. At $t = 0$, the velocity is 8 ms^{-1}.

       (b)   The ball reaches its maximum height when its velocity is zero.

$$8 - 10t = 0 \Rightarrow t = 0.8$$

          At $t = 0.8$, $h = 5.2$

          The ball reaches its maximum height of 5.2 metres after 0.8 seconds.

       (c)   The acceleration is $\frac{d^2h}{dt^2} = -10$, which is constant. 10 ms^{-2} is the approximate value of the acceleration due to gravity. It is negative because the acceleration is downwards.

(continued)

5. (a) $x = 3 \cos 2t$ $y = 3 \sin 2t$

    $\dfrac{dx}{dt} = -6 \sin 2t$ $\dfrac{dy}{dt} = 6 \cos 2t$

    $\dfrac{d^2x}{dt^2} = -12 \cos 2t$ $\dfrac{d^2y}{dt^2} = -12 \sin 2t$

    After 1 second; $\dfrac{dx}{dt} = -5.46 \text{ ms}^{-1}$, $\dfrac{dy}{dt} = -2.50 \text{ ms}^{-1}$

    $\dfrac{d^2x}{dt^2} = 4.99 \text{ ms}^{-2}$, $\dfrac{d^2y}{dt^2} = -10.91 \text{ ms}^{-2}$

    After 2 seconds; $\dfrac{dx}{dt} = 4.54 \text{ ms}^{-1}$, $\dfrac{dy}{dt} = -3.92 \text{ ms}^{-1}$

    $\dfrac{d^2x}{dt^2} = 7.84 \text{ ms}^{-2}$, $\dfrac{d^2y}{dt^2} = 9.08 \text{ ms}^{-2}$

    (b) $\dfrac{dx}{dt} = -6 \sin 2t = -2(3 \sin 2t) = -2y$

    $\dfrac{d^2x}{dt^2} = -12\cos 2t = -4(3 \cos 2t) = -4x$

    (c) $\dfrac{dy}{dt} = 6 \cos 2t = 2(3 \cos 2t) = 2x$

    $\dfrac{d^2y}{dt^2} = -12 \sin 2t = -4(3 \sin 2t) = -4y$

    (d)   At a point on the circle with position vector $\begin{bmatrix} x \\ y \end{bmatrix}$, the velocity vector is $\begin{bmatrix} -2y \\ 2x \end{bmatrix}$

    and the acceleration vector is $\begin{bmatrix} -4x \\ -4y \end{bmatrix}$.

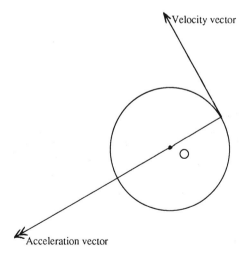

Velocity vector

Acceleration vector

(continued)

13

6.   (a)

	(i)	(ii)	(iii)	(iv)	(v)	(vi)
x	y	y	y	y	y	y
−2.0	75.0	50.0	25.0	0.0	−25.0	−50.0
−1.6	28.5	18.9	9.3	−0.3	−9.9	−19.5
−1.2	13.6	8.9	4.2	−0.5	−5.2	−9.9
−0.8	7.9	5.0	2.2	−0.7	−3.5	−6.4
−0.4	5.5	3.4	1.2	−0.9	−3.0	−5.1
0.0	4.4	2.6	0.7	−1.2	−3.0	−4.9
0.4	4.2	2.3	0.3	−1.6	−3.6	−5.5
0.8	4.7	2.3	0.0	−2.4	−4.7	−7.1
1.2	6.1	2.8	−0.5	−3.7	−7.0	−10.3
1.6	9.2	4.0	−1.2	−6.4	−11.6	−16.8
2.0	16.0	6.6	−2.7	−12.0	−21.3	−30.6

(b)

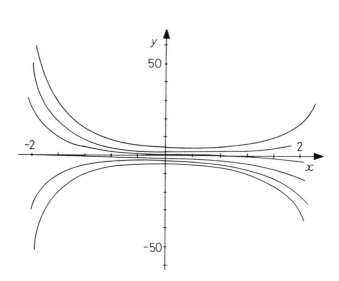

(c)   No matter what starting point is chosen, the following values result:

x	−2.0	−1.5	−1.0	−0.5	0.0	0.5	1.0	1.5	2.0
y	−	−0.5	−0.6	−0.8	−1.1	−1.6	−2.5	−4.3	−8.0

This happens because, for $x = -2$ and $dx = 0.5$,

$$\frac{dy}{dx} = xy - 1$$
$$\Rightarrow \quad dy = (-2y - 1)\,0.5$$
$$\Rightarrow y + dy = -0.5 \text{ for any initial } y \text{ value.}$$

Because of a freak coincidence in the choice of step, a solution curve has not been obtained.

2 Numerical solutions

2.2 Simultaneous linear equations

In a chemical reaction involving only butane and methane, the rate at which butane changes into methane depends on the quantity of each present.

Under certain physical conditions, the rates of change are given by

$$\frac{dx}{dt} = -0.2x + 0.7y \quad \text{and} \quad \frac{dy}{dt} = 0.2x - 0.7y$$

where x and y, respectively, denote the overall percentages of butane and methane in the mixture after time t minutes.

(a) Write a program to estimate the percentages of butane and methane after 3 minutes if there are equal quantities of each when the reaction starts.

(b) Explain why

    (i) $x + y = 100$ (ii) $\dfrac{dy}{dt} = -\dfrac{dx}{dt}$

(a) Two possible programs are:

BASIC

```
10  INPUT t, x, y
20  INPUT dt
30  PRINT t, x, y
40  dx = (–0.2 * x + 0.7 *y)*dt
50  dy = (0.2 *x – 0.7*y)*dt
60  PRINT dt, dx, dy
70  t = t + dt
80  x = x + dx
90  y = y + dy
100 GOTO 30
```

fx 7700 GA

```
"T = " ? → T
"X = " ? → X
"Y = " ? → Y
"D = " ? → D
Lbl 1
(– 0.2X + 0.7Y) x D → E
(0.2X – 0.7Y) x D → F
"DT = " : D ◢
"DX = " : E ◢
"DY = " : F ◢
T + D → T : "T = " : T ◢
X + E → X : "X = " : X ◢
Y + F → Y : "Y = " : Y ◢
GOTO 1
```

Once a program has been written and tested, it is easy to investigate changing both the initial conditions and the size of the step.

For a step of $dt = 0.5$, the following results should be obtained.

t	x	y	dt	dx	dy
0	50.0	50.0	0.5	12.5	−12.5
0.5	62.5	37.5	0.5	6.9	−6.9
1.0	69.4	30.6	0.5	3.8	−3.8
1.5	73.2	26.8	0.5	2.1	−2.1
2.0	75.2	24.8	0.5	1.1	−1.1
2.5	76.4	23.6	0.5	0.6	−0.6
3.0	77.0	23.0			

It is estimated that, after 3 minutes, the percentages are

Butane 23%
Methane 77%

(b) (i) Only butane and methane are involved in this reaction and so the total quantity of the two gases is always 100%.

(ii) Since $x + y$ is constant, the rate at which x increases is the same as the rate at which y decreases. So

$$\frac{dy}{dt} = -\frac{dx}{dt}$$

Stereo tuners

1. $$\frac{di}{dt} = 3 - 4i - \frac{q}{10}, \qquad \frac{dq}{dt} = i$$

t	i	q	dt	di	dq
0	0	0	0.2	0.6	0
0.2	0.60	0	0.2	0.12	0.12
0.4	0.72	0.12	0.2	0.02	0.14
0.6	0.74	0.26	0.2	0.00	0.15
0.8	0.74	0.41	0.2	0.00	0.15
1.0	0.74	0.56			

2. (a) After an initial surge, the charge steadily increases. The current quickly rises to a maximum, which is reached after about 1 minute; it then falls off slowly.

    (b) The initial conditions affect the values of the current and charge but not their general behaviour over time. The current adjusts itself within 1 minute to a value from which it slowly decreases, whilst the charge steadily increases.

3. $$\frac{di}{dt} = 3\cos t - 4i - \frac{q}{10}, \qquad \frac{dq}{dt} = i$$

    The results tabulated are for a step of $dt = 0.1$.

t	0	1	2	3	4	5
i	0	0.52	−0.15	−0.69	−0.60	0.05
q	0	0.51	0.75	0.32	−0.39	−0.73

Both the current and the charge appear to oscillate. This is confirmed by extending the time period.

t	5	6	7	8	9	10
i	0.05	0.65	0.65	0.06	−0.59	−0.70
q	−0.73	−0.38	0.33	0.75	0.49	−0.21

Changing the initial conditions does not affect the oscillatory nature, which can be seen in both current and charge after a short settling down period.

Second derivatives

1.   (a)

t	0	1	2	3	4	5
x	10	14	8	−12	−40	−56
v	4	−6	−20	−28	−16	24

When $t = 5$, $x \approx -56$ and $y \approx 24$.

(b)

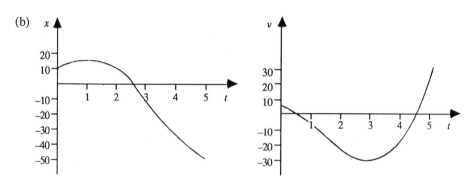

(c)

t	0	1	2	3	4	5
x	10	9.2	−0.5	−10.8	−11.7	−1.5
v	4	−6.5	−11.9	−6.34	5.9	13.7

When $t = 5$, $x \approx -1.5$ and $y \approx 13.7$

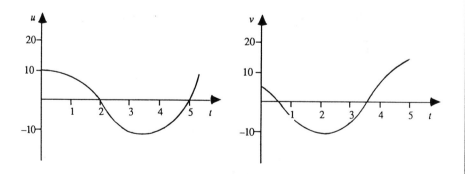

This solution suggests that using d$t = 1$ gives a very inaccurate estimate.

(continued)

18

(d) $\dfrac{d^2x}{dt^2} = -x$ or, more generally, $\dfrac{d^2x}{dt^2} = -kx$

is the differential equation that represents simple harmonic motion (SHM).

Two common examples of SHM are:

- a weight oscillating on the end of a spring;

- a simple pendulum swinging through a small angle of displacement.

2E.   (a) $dx = v\,dt$, $dv = y\,dt$ and $dy = (2t - 6x - 11v - 6y)dt$

t	0	0.5	1.0	1.5	2.0	2.5	3.0	3.5	4.0
x	2	3	4.5	0.75	1.25	0.44	0.69	0.61	0.80
v	2	3	−7.5	1	−1.63	0.5	−0.16	0.38	0.21
y	2	−21	17	−5.25	4.25	−1.31	1.06	−0.33	0.27

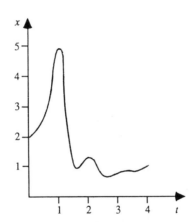

(b) $dx = v\,dt$, $dv = y\,dt$ and $dy = (e^t - 7x + 5t^2v - 3ty)dt$

t	1	1.2	1.4	1.6	1.8	2.0
x	4	4.4	4.8	5.1	5.2	5.3
v	2	2.0	1.6	1.0	0.3	−0.5
y	1	−1.9	−2.9	−3.3	−3.9	−5.0

When $t = 2$, $x \approx 5.3$, $\dfrac{dx}{dt} \approx -0.5$ and $\dfrac{d^2x}{dt^2} \approx -5.0$

Tutorial sheet

1. For a step of $dt = 0.1$:

t	0	1	2	3	4	5	10
i	0	0.7	0.62	0.54	0.48	0.42	0.22
q	0	0.55	1.21	1.80	2.31	2.76	4.31

After a surge, the current appears to gradually die away.

The charge appears to be steadily increasing. Putting $i = 0$ in the differential equations gives

$$0 = 3 - \frac{q}{2} \quad \text{and} \quad \frac{dq}{dt} = 0$$

These equations are solved by $q = 6$. It therefore seems likely that i tends to 0 and q tends to 6.

2. Let $v = \frac{dy}{dt}$, then

$$\frac{dv}{dt} = 2v - \frac{5000}{y^2}$$

Using $dt = 0.1$ the following results are obtained,

t	0	0.5	1	1.5	2	2.5	3
y	100	99.9	99.6	98.6	95.8	88.8	71.0
v	0	−0.4	−1.3	−3.6	−9.4	−23.7	−59.6

At $t = 3$, $y \approx 71$ and $v \approx 60$ downwards.

3. $dx = 4t\, dt, \quad dy = 6t\, dt, \quad dz = 8 \sin 0.2t\, dt$

t	0	1	2	3	4	5	6	7	8	9	10
x	0	1	6	15	28	45	66	91	120	153	190
y	0	1.5	9	22.5	42	67.5	99	136.5	180	229.5	285
z	100	100.4	102.4	105.9	110.7	116.7	123.6	131.2	139.1	147.1	154.8

(a) The estimated position at $t = 2$ is $(6, 9, 102)$.

(continued)

(b)

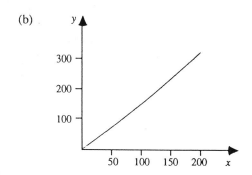

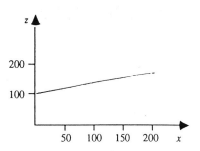

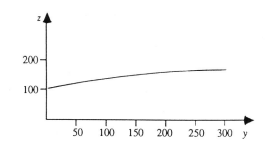

4E. $dx = \dfrac{x}{y}\, dt,\quad dy = \dfrac{y}{z}\, dt,\quad dz = \dfrac{z}{x}\, dt$

(a)

t	0	0.5	1.0	1.5	2.0	2.5	3.0
x	1	1.5	2.0	2.5	3.0	3.5	4.0
y	1	1.5	2.0	2.5	3.0	3.5	4.0
z	1	1.5	2.0	2.5	3.0	3.5	4.0

(b)

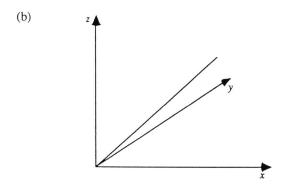

For these particular initial conditions, the path appears to be a straight line. It may be of interest to investigate other initial conditions.

(continued)

5E. A step of dt = 0.5 gives:

t	0	0.5	1.0	1.5	2.0
x	1	1	71	−176	7050
y	1	15	21.6	1269	−69070
z	1	0.17	7.4	764	−11986

This solution is very erratic in behaviour, and you can have no confidence in its validity.

Using dt = 01:

t	0	0.2	0.4	0.6	0.8	1.0
x	1	3.8	16.7	99.3	4.56×10^3	8.88×10^7
y	1	6.3	34.6	419.1	1.92×10^5	1.68×10^{12}
z	1	0.99	12.8	393.2	1.94×10^5	1.70×10^{12}

The rate of growth of x, y and z suggests these solutions are likely to be very inaccurate. Since these differential equations are connected with chaos theory, it is perhaps not surprising that a numerical solution behaves erratically.

3 First order equations

3.1 Separable variables

> **(a)** Explain why any circle, centre the origin, has an equation of the form $x^2 + y^2 = c$.
>
> **(b)** From the equation $x^2 + y^2 = c$, use implicit differentiation to obtain $\frac{dy}{dx} = -\frac{x}{y}$.
>
> **(c)** Reverse the process in (b) to show how $x^2 + y^2 = c$ can be obtained from $\frac{dy}{dx} = -\frac{x}{y}$.

(a) Let r be the radius of the circle and let P be any point on the circle. From Pythagoras' Theorem,

$$x^2 + y^2 = r^2$$
$$\Rightarrow x^2 + y^2 = \text{constant}$$

(b)
$$x^2 + y^2 = \text{constant}$$
$$\Rightarrow 2x + 2y\frac{dy}{dx} = 0$$
$$\Rightarrow \qquad \frac{dy}{dx} = -\frac{x}{y}$$

(c) The reverse of the argument in (b) is:

$$\frac{dy}{dx} = -\frac{x}{y}$$
$$\Rightarrow y\frac{dy}{dx} = -x$$
$$\Rightarrow \frac{1}{2}y^2 = -\frac{1}{2}x^2 + C \qquad \text{[Integrating with respect to } x\text{]}$$
$$\Rightarrow x^2 + y^2 = c \qquad\qquad [c = 2C]$$

It is simple to think of this method in terms of the symbols dx and dy :

$$\frac{dy}{dx} = -\frac{x}{y}$$
$$\Rightarrow \quad y\,dy = -x\,dx \qquad ①$$
$$\Rightarrow \quad \int y\,dy = -\int x\,dx$$
$$\Rightarrow \quad \frac{1}{2}y^2 = -\frac{1}{2}x^2 + C$$
$$\Rightarrow x^2 + y^2 = c$$

Equation ① is the crucial stage of the argument. The terms involving x and the terms involving y are 'separated' by being put on opposite sides of the equation.

Particular integrals

1. $-2.5C = 7.4$
    $C = -2.96$

2. $B + 0.01A + 0.01Bt = 0.025t$

    Equating coefficients of t,
    $$0.01B = 0.025 \Rightarrow B = 2.5$$

    Equating constant coefficients,
    $$B + 0.01A = 0 \Rightarrow A = -250$$

3. $B + 0.01A + 0.01Bt = 0.5 - 0.025t$
    $B = -2.5$
    $A = 300$

4. (a) Try $y = A + Bt + Ct^2$

    $B + 2Ct + 0.01 (A + Bt + Ct^2) = 0.0001t^2 + 0.03t - 1$

$0.01C = 0.0001$	Coefficients of t^2
$2C + 0.01B = 0.03$	Coefficients of t
$B + 0.01A = -1$	Constant coefficients

    $C = 0.01$
    $B = 1$
    $A = -200$

    (b) Try $y = Ae^{0.02t}$

    $$0.02Ae^{0.02t} + 0.01Ae^{0.02t} = 0.3e^{0.02t}$$
    $\Rightarrow \quad 0.03A = 0.3$
    $\Rightarrow \qquad A = 10$

Linearity

FIRST ORDER EQUATIONS
COMMENTARY
TASKSHEET 2

1. (a) $\cos x - 3 \sin x$ (b) $-\dfrac{1}{x^2} - \dfrac{3}{x}$ (c) $\cos x - \dfrac{1}{x^2} - 3 \sin x - \dfrac{3}{x}$

2. (a) 20 (b) $1 + 5x$ (c) $1 + 20 + 5x = 21 + 5x$

3. (a) $2x + x^4$ (b) $5 + 5x^3$

   (c) $e^x + x^2 e^x$ (d) $2x + 5 + e^x + x^4 + 5x^3 + x^2 e^x$

4. (a) 16 (b) $1 + x^2$ (c) $1 + (4 + x)^2 = 17 + 8x + x^2$

5. In question 4, the answer to (c) is **not** the sum of the answers to (a) and (b). The expressions in questions 1 to 3 **are** linear in y and $\dfrac{dy}{dx}$ and so there are additive relationships between the various answers.

6E. $\dfrac{dy}{dx} + p(x)y = \dfrac{dy_1}{dx} + \dfrac{dy_2}{dx} + p(x)(y_1 + y_2)$

$\qquad\qquad = \dfrac{dy_1}{dx} + p(x)y_1 + \dfrac{dy_2}{dx} + p(x)y_2$

If y_1 is the complementary function, then it is the general solution of $\dfrac{dy}{dx} + p(x)y = 0$.

If y_2 is a particular integral, then it is a solution of $\dfrac{dy}{dx} + p(x)y = q(x)$.

For $y = y_1 + y_2$,

$\qquad \dfrac{dy}{dx} + p(x)y = \dfrac{dy_1}{dx} + p(x)y_1 + \dfrac{dy_2}{dx} + p(x)y_2$

$\qquad\qquad\qquad = 0 + q(x)$

$\qquad\qquad\qquad = q(x)$

$y_1 + y_2$ is therefore the general solution of the differential equation.

Integrating factors

1. (a) $\dfrac{d}{dx}\left(e^{3x}y\right) = 7e^{3x}$

   $e^{3x}y = \dfrac{7}{3}e^{3x} + c$

   $y = \dfrac{7}{3} + ce^{-3x}$

   (b) $\dfrac{d}{dx}\left(e^{-2x}y\right) = 9e^{-2x}$

   $e^{-2x}y = -\dfrac{9}{2}e^{-2x} + c$

   $y = -\dfrac{9}{2} + ce^{2x}$

   (c) $\dfrac{d}{dx}\left(e^{x}y\right) = xe^{x}$

   $e^{x}y = \displaystyle\int xe^{x}\,dx$

   $e^{x}y = (x-1)e^{x} + c$

   $y = x - 1 + ce^{-x}$

2. (a) e^{kx}

   (b) In 1(c), note that the function xe^{x} had to be integrated. In general, the integration required may be difficult or even impossible.

3. $\displaystyle\int \dfrac{3}{x}\,dx = 3\ln x, \ \ IF = x^{3}$

   $\dfrac{d}{dx}\left(x^{3}y\right) = x^{2}$

   $y = \dfrac{1}{3} + \dfrac{c}{x^{3}}$

4. (a) $\dfrac{d}{dx}\left(ye^{x}\right) = 1$

   $y = (x+c)e^{-x}$

   $y = (x+1)e^{-x}$

   (b) $\dfrac{d}{dx}\left(ye^{x^{2}}\right) = xe^{x^{2}}$

   $y = 1 + ce^{-x^{2}}$

   $y = 1 + 2e^{-x^{2}}$

   (c) $IF = \cos x$

   $\dfrac{d}{dx}\left(y\cos x\right) = \cos x$

   $y\cos x = -\sin x + c$

   $y = -\tan x + c\sec x$

   $y = \dfrac{1}{\sqrt{2}}\sec x - \tan x$

Tutorial sheet

1. (a) If no water were being drained out, the rate of increase of the bacteria would be of the form kN, for some constant k.

Since water is being drained at a rate of 2 litres per hour, this represents a loss of bacteria at a rate of $\dfrac{2}{100-2t} \times N$ per hour.

$\dfrac{dN}{dt}$ is therefore $kN - \dfrac{2}{100-2t} \times N$

(b) $\dfrac{dN}{dt} = \left(k - \dfrac{2}{100-2t} \right) N$

$\Rightarrow \displaystyle\int \dfrac{dN}{N} = \int \left(k - \dfrac{2}{100-2t} \right) dt$

$\Rightarrow \ln N = kt + \ln(100 - 2t) + c$

$\Rightarrow N = e^{kt + \ln(100 - 2t) + c}$

$\Rightarrow N = \dfrac{100 - 2t}{100} N_0 e^{kt}$

(c) $\dfrac{dN}{dt} = \left(k - \dfrac{2}{100-2t} \right) N$

If $k \le \dfrac{1}{50}$, then the number of bacteria decreases as the water drains out. After 50 hours, the tank (theoretically) contains no water and no bacteria.

If $k > \dfrac{1}{50}$, then the number of bacteria initially increases. After time $50 - \dfrac{1}{k}$, the number decreases. The number is again zero after 50 hours.

2. (a) $\dfrac{dy}{dx} = -\dfrac{2A}{x^3}$

$x\dfrac{dy}{dx} = -2y$

(b) $2y\dfrac{dy}{dx} = 1$

(c) $4\dfrac{dv}{dt} = \dfrac{2}{t}(\ln t + c)$

$t\dfrac{dv}{dt} = \sqrt{(v+1)}$

(continued)

3. (a) $y = 0.875 + 1.5x$

(b) $x = 1.75e^t$

(c) $y = -0.78125 - 0.125x - 0.25x^2$

4. (a) $q = Ae^{-t} + \cos t + \sin t$

(b) $q = Ae^{-t} + \sin t - \cos t$

(c) $q = Ae^{-t} + \frac{1}{2}(\cos t + \sin t) + \frac{1}{2}(\sin t - \cos t)$

$= Ae^{-t} + \sin t$

5E. (a) $\frac{dq}{dt} + \frac{q}{RC} = 0$

$q = Ae^{-\frac{1}{RC}t}$

The charge tends to zero.

(b) $\text{CF} = Ae^{-\frac{1}{RC}t}, \quad \text{PI} = CE_0$

$q = Ae^{-\frac{1}{RC}t} + CE_0$

The charge tends to CE_0.

(c) $\text{CF} = Ae^{-\frac{1}{RC}t}$

Let $\text{PI} = \alpha \cos pt + \beta \sin pt$

$Rp(-\alpha \sin pt + \beta \cos pt) + \frac{\alpha}{c}\cos pt + \frac{\beta}{c}\sin pt = E_0 \cos pt$

$Rp\beta + \frac{\alpha}{c} = E_0$

$-Rp\alpha + \frac{\beta}{c} = 0$

$\alpha = \frac{CE_0}{1 + C^2 R^2 p^2}$

$\beta = \frac{C^2 Rp E_0}{1 + C^2 R^2 p^2}$

$q = Ae^{-\frac{1}{RC}t} + \frac{CE_0}{1 + C^2 R^2 p^2}(\cos pt + CRp \sin pt)$

As $t \to +\infty$, q oscillates sinusoidally.

4 Second order equations

4.1 Arbitrary constants

> (a)   Obtain the general solution of $\dfrac{d^2y}{dt^2} = -10$.
>
> (b)   How many arbitrary constants does your solution possess? In general, explain how and why the number of arbitrary constants is related to the order of a differential equation.
>
> (c)   For the differential equation $\dfrac{d^2y}{dt^2} = -10$, describe possible initial conditions that would determine the values of the arbitrary constants and thus give rise to a particular solution. How many initial conditions would be needed to determine the arbitrary constants for a differential equation of order n?

(a)   The differential equation can be integrated twice.

$$\frac{d^2y}{dt^2} = -10$$

$$\Rightarrow \frac{dy}{dt} = -10t + a$$

$$\Rightarrow y = -5t^2 + at + b$$

(b)   The general solution has two arbitrary constants. The number of arbitrary constants is the same as the order of the differential equation.

Solving a differential equation involves performing integration, the inverse of differentiation. Each time this is done, an arbitrary constant is introduced. In solving an nth order differential equation, there are n stages of integration and so there will be n arbitrary constants in the general solution.

(c)   Perhaps the most likely initial conditions are

-     the initial height (this determines b);

-     the initial speed (this determines a).

In general, n conditions are needed to determine n constants. However, the conditions need not be 'initial' ones and are, more generally, called *boundary conditions*. In the case of a falling body, other possible conditions include the times taken to fall particular distances or the speeds at specified times.

Adding solutions

1. (a) $9e^{-3x} - 3e^{-3x} - 6e^{3x} = 0$

    (b) $9Ae^{-3x} - 3Ae^{-3x} - 6Ae^{-3x} = 0$

    (c) $4Be^{2x} + 2Be^{2x} - 6Be^{2x} = 0$

    (d) $(9Ae^{-3x} + 4Be^{2x}) + (-3Ae^{-3x} + 2Be^{2x}) + (-6Ae^{-3x} - 6Be^{2x}) = 0$

    Result (d) demonstrates the linearity of the differential equation.

2. (a) $y = x^2 \Rightarrow \frac{dy}{dx} = 2x$ and $\frac{d^2y}{dx^2} = 2$

    Then $\frac{d^2y}{dx^2} + \frac{dy}{dx} - 6y = 2 + 2x - 6x^2$

    (b) This is the same as the 'CF + PI' result for first order linear equations.

3. (a) –
    (b) –
    (c) The 'CF + PI' result holds because of the linearity of $\frac{d^2y}{dx^2} + 3\frac{dy}{dx} + 2y$.

4. $y = Ae^{-x} + Be^{-2x} + \sin 2x$

5. (a) $10x^2 - 14x + 2$
    (b) $4e^x$
    (c) $10x^2 - 14x + 2 + 4e^x$
    (d) 0
    (e) $10x^2 - 14x + 2$

6E. (a) $\frac{d^2y}{dx^2} - 5\frac{dy}{dx} + 6y = \frac{d^2y_1}{dx^2} + \frac{d^2y_2}{dx^2} - 5\left(\frac{dy_1}{dx} + \frac{dy_2}{dx}\right) + 6(y_1 + y_2)$

    $= \frac{d^2y_1}{dx^2} - 5\frac{dy_1}{dx} + 6y_1 + \frac{d^2y_2}{dx^2} - 5\frac{dy_2}{dx} + 6y_2$

    $= 0$

    (b) $\frac{d^2y}{dx^2} - 5\frac{dy}{dx} + 6y$

    $= A\left(\frac{d^2y_1}{dx^2} - 5\frac{dy_1}{dx} + 6y_1\right) + B\left(\frac{d^2y_2}{dx^2} - 5\frac{dy_2}{dx} + 6y_2\right) + \frac{d^2y_3}{dx^2} - 5\frac{dy_3}{dx} + 6y_3$

    $= 1$

Complementary functions

1. $\dfrac{dy}{dx} = me^{mx}$, $\dfrac{d^2y}{dx^2} = m^2 e^{mx}$

2. $y = e^{mx}$ satisfies $\dfrac{d^2y}{dx^2} - 7\dfrac{dy}{dx} + 10y = 0$

    $\Leftrightarrow (m^2 - 7m + 10)\,e^{mx} = 0$

    $\Leftrightarrow \quad m^2 - 7m + 10 = 0$

    $\Leftrightarrow \quad m = 2 \text{ or } m = 5$

    Then $Ae^{2x} + Be^{5x}$ is a solution. Since it contains two arbitrary constants, it is the general solution.

3. $y = e^{mx}$ satisfies $\dfrac{d^2y}{dx^2} - 5\dfrac{dy}{dx} + 6y = 0$

    $\Leftrightarrow (m^2 - 5m + 6)\,e^{mx} = 0$

    $\Leftrightarrow \quad m^2 - 5m + 6 = 0$

    $\Leftrightarrow \quad m = 2 \text{ or } m = 3$

    The complementary function is therefore
    $$y = Ae^{2x} + Be^{3x}$$
    When $x = 0$, $y = A + B = 1$
    $$\dfrac{dy}{dx} = 2A + 3B = 2$$

    Therefore, $A = 1$ and $B = 0$. The required solution is $y = e^{2x}$.

4. (a) $m^2 + 5m + 4 = 0$

    (b) $m^2 - m - 12 = 0$

    (c) $m^2 + 2m + 1 = 0$

    (d) $m^2 + 1 = 0$

    (e) $m^2 - 8 = 0$

    (f) $am^2 + bm + c = 0$

Particular integrals

1. $\dfrac{dy}{dx} = a, \ \dfrac{d^2y}{dx^2} = 0$

$-5a + 6(ax + b) = x$

$\Rightarrow 6ax + (6b - 5a) = x$

$\Rightarrow \quad a = \dfrac{1}{6}, \ b = \dfrac{5}{36}$

The particular integral is $\dfrac{1}{6}x + \dfrac{5}{36}$.

2. (a) $10k = 3 \ \Rightarrow k = 0.3$

$y = 0.3$

(b) $10ax + 10b - 7a = 20x - 64 \Rightarrow a = 2, b = -5$

$y = 2x - 5$

(c) $10ax^2 + (10b - 14a)x + (10c - 7b + 2a) = 10x^2 - 14x + 22$

$\Rightarrow a = 1, b = 0, c = 2$

$y = x^2 + 2$

(d) $C - 7C + 10C = 8 \ \Rightarrow C = 2$

$y = 2e^x$

(e) $9C - 21C + 10C = 1 \ \Rightarrow C = -\dfrac{1}{2}$

$y = -\dfrac{1}{2} e^{3x}$

(f) $(-a + 7b + 10a)\sin x + (-b - 7a + 10b)\cos x = \sin x$

$\Rightarrow a = \dfrac{9}{130} , \ b = \dfrac{7}{130}$

$y = \dfrac{1}{130} (9\sin x + 7\cos x)$

3. $(-a + 4a)\sin x + (-b + 4b)\cos x = 3\sin x + 6\cos x$

$\Rightarrow a = 1, \ b = 2$

$y = \sin x + 2\cos x$

1.   (a)

$$y = \cos x \qquad\qquad\qquad y = \sin x$$

$$\Rightarrow \frac{dy}{dx} = -\sin x \qquad\qquad \Rightarrow \frac{dy}{dx} = \cos x$$

$$\Rightarrow \frac{d^2y}{dx^2} = -\cos x \qquad\qquad \Rightarrow \frac{d^2y}{dx^2} = -\sin x$$

   (b)   Both $y = \cos x$ and $y = \sin x$ satisfy the differential equation

$$\frac{d^2y}{dx^2} = -y$$

i.e. $\dfrac{d^2y}{dx^2} + y = 0$

   (c)

$$y = A \cos x + B \sin x$$

$$\Rightarrow \frac{dy}{dx} = -A \sin x + B \cos x$$

$$\Rightarrow \frac{d^2y}{dx^2} = -A \cos x - B \sin x$$

$$\Rightarrow \frac{d^2y}{dx^2} = -y \quad \text{i.e.} \quad \frac{d^2y}{dx^2} + y = 0$$

This is the **general** solution because it has two arbitrary constants.

2.   (a)   The auxiliary equation is $m^2 + 1 = 0$

$$\Leftrightarrow m = \pm j$$

The complementary function is therefore $Ce^{jx} + De^{-jx}$

   (b) $Ce^{jx} + De^{-jx} = C(\cos x + j \sin x) + D(\cos x - j \sin x)$

$$= (C + D)\cos x + (Cj - Dj)\sin x$$

[For the CF to be real, C and D must be complex numbers such that $C + D$ and $Cj - Dj$ are both real.]

(continued)

3.   (a) $y = \cos 3x$ $y = \sin 3x$

$$\Rightarrow \frac{dy}{dx} = -3 \sin 3x \qquad\qquad \Rightarrow \frac{dy}{dx} = 3 \cos 3x$$

$$\Rightarrow \frac{d^2 y}{dx^2} = -9 \cos 3x \qquad\qquad \Rightarrow \frac{d^2 y}{dx^2} = -9 \sin 3x$$

(b) $y = A \cos 3x + B \sin 3x$

By linearity, $\dfrac{d^2 y}{dx^2} + 9y = A(-9 \cos 3x + 9 \cos 3x) + B(-9 \sin 3x + 9 \sin 3x)$

$$= 0$$

4.   (a) $A \cos 4x + B \sin 4x$

       (b) $A \cos 2t + B \sin 2t$

       (c) $A \cos \omega x + B \sin \omega x$

       (d) $A \cos \sqrt{2}\, v + B \sin \sqrt{2}\, v$

       (e) $A \cos \sqrt{5}\, t + B \sin \sqrt{5}\, t$

5. $y = A \cos 3x + B \sin 3x$

$$\frac{dy}{dx} = -3A \sin 3x + 3B \cos 3x$$

When $x = 0$, $0 = A$

                      $1 = 3B$

$$y = \frac{1}{3} \sin 3x$$

6.   (a)   (i) $\dfrac{dy}{dx} = e^x \dfrac{dz}{dx} + e^x z$

$$\frac{d^2 y}{dx^2} = e^x \frac{d^2 z}{dx^2} + 2e^x \frac{dz}{dx} + e^x z$$

$$e^x \left[\frac{d^2 z}{dx^2} + 2\frac{dz}{dx} + z - 2\left(\frac{dz}{dx} + z \right) + 10z \right] = 0$$

$$\Rightarrow \frac{d^2 z}{dx^2} + 9z = 0$$

$$\Rightarrow z = A \cos 3x + B \sin 3x$$

$$\Rightarrow y = e^x (A \cos 3x + B \sin 3x)$$

(continued)

(ii) $m^2 - 2m + 10 = 0$

$\Rightarrow m = \dfrac{2 \pm \sqrt{(4-40)}}{2}$

$\Rightarrow m = 1 \pm 3j$

(b) $e^{\upsilon x}(A \cos \omega x + B \sin \omega x)$

This result is proved in question 7E.

7E.   (a)   (i) $\dfrac{dy}{dx} = e^{\upsilon x} \dfrac{dz}{dx} + \upsilon e^{\upsilon x} z$

$\dfrac{d^2 y}{dx^2} = e^{\upsilon x} \dfrac{d^2 z}{dx^2} + 2\upsilon e^{\upsilon x} \dfrac{dz}{dx} + \upsilon^2 e^{\upsilon x} z$

$e^{\upsilon x} \left[\dfrac{d^2 z}{dx^2} + 2\upsilon \dfrac{dz}{dx} + \upsilon^2 z - 2\upsilon \left(\dfrac{dz}{dx} + \upsilon z \right) + (\upsilon^2 + \omega^2) z \right] = 0$

$\Rightarrow e^{\upsilon x} \left(\dfrac{d^2 z}{dx^2} + \omega^2 z \right) = 0$

$\Rightarrow \quad \dfrac{d^2 z}{dx^2} + \omega^2 z = 0$

$\Rightarrow \quad z = A \cos \omega x + B \sin \omega x$

$\Rightarrow \quad y = e^{\upsilon x}(A \cos \omega x + B \sin \omega x)$

(ii) $m^2 - 2\upsilon m + (\upsilon^2 + \omega^2) = 0$

$\Rightarrow m = \dfrac{2\upsilon \pm \sqrt{(4\upsilon^2 - 4(\upsilon^2 + \omega^2))}}{2}$

$\Rightarrow m = \upsilon \pm \omega j$

(b)   A second order differential equation whose auxiliary equation has roots $\upsilon \pm \omega j$ has a complementary function given by

$e^{\upsilon x}(A \cos \omega x + B \sin \omega x)$

1.   (a) $\dfrac{d^2x}{dt^2}$ is the acceleration of the object. $\dfrac{d^2x}{dt^2}$ is $-4x$ and so the acceleration has

   magnitude 4 times the displacement of the object from its equilibrium position. The acceleration is directed **towards** the equilibrium position.

   (b)   The general solution is

   $$x = A \cos 2t + B \sin 2t$$
   $$\frac{dx}{dt} = -2A \sin 2t + 2B \cos 2t$$

   Then $0.1 = A$ and $0.2 = 2B$ and so

   $$x = 0.1 \cos 2t + 0.1 \sin 2t$$

   (c) $x = C \sin (2t + \varepsilon) = (C \cos \varepsilon) \sin 2t + (C \sin \varepsilon) \cos 2t$

   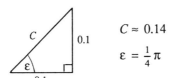

   $C \approx 0.14$

   $\varepsilon = \dfrac{1}{4} \pi$

   $$x \approx 0.14 \sin \left(2t + \frac{\pi}{4}\right)$$

   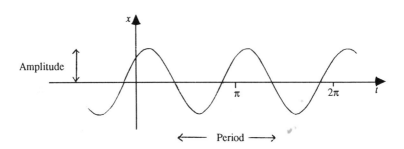

   $C \approx 0.14$ is the amplitude of the oscillations. $\dfrac{2\pi}{\omega} = \pi$ is the period of the oscillations.

2.   (a)   The auxiliary equation is $m^2 + 5m + 4 = 0$.

   $$x = Ae^{-t} + Be^{-4t}$$

   (b)   For **any** initial conditions, x tends to zero exponentially as t increases.

   (continued)

3.   (a)   The auxiliary equation is $m^2 + 6m + 13 = 0$

$$\Leftrightarrow m = -3 \pm 2j$$

$x = e^{-3t}(A \cos 2t + B \sin 2t)$

$x = 0$ when $t = 0 \Rightarrow A = 0$ and $x = Be^{-3t} \sin 2t$

Then $\dfrac{dx}{dt} = -3Be^{-3t} \sin 2t + 2Be^{-3t} \cos 2t$

$\Rightarrow 0.2 = 2B$

$\Rightarrow \quad x = 0.1\, e^{-3t} \sin 2t$

(b)

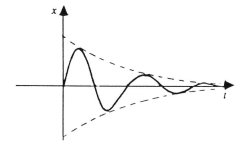

The solution again decays exponentially. In this case, the solution is still oscillatory in nature.

4.   (a)   Try a PI of the form $A \cos 2t + B \sin 2t$

$(-4A + 12B + 13A) \cos 2t + (-4B - 12A + 13B) \sin 2t = 10 \sin 2t$

$9A + 12B = 0$ and $9B - 12A = 10$

$A = -\dfrac{8}{15}, \ B = \dfrac{2}{5}$

$\text{PI} = \dfrac{2}{5} \sin 2t - \dfrac{8}{15} \cos 2t$

$x = e^{-3t}(A \cos 2t + B \sin 2t) + \dfrac{2}{5} \sin 2t - \dfrac{8}{15} \cos 2t$

(b)   As t increases, the CF tends to zero and so x tends to the sinusoidal PI.

The CF is often called the **free vibration** of the system whereas the PI is called the **forced vibration**. You might like to investigate cases where the free and forced vibrations have the same frequency. In such cases, the system is said to **resonate.** You should be able to show why resonance can lead to damage in machines and vehicles which vibrate when in use.